Bibliographic information published by the German National Library:

The German National Library lists this publication in the National Bibliography; detailed bibliographic data are available on the Internet at http://dnb.dnb.de .

Imprint:

Print and binding: Books on Demand GmbH, Norderstedt Germany
ISBN: 9783346198402

This book at GRIN:

https://www.grin.com/document/541268

Mirroyal Ismayilov

Predictive Control Algorithms in Human Balancing

GRIN Verlag

TDK

Predictive control algorithms in human balancing

Ismayilov Mirroyal

14 November 2018

Budapest University of Technology and Economics

Acknowledgement

I would like to thank Dr. Zelei Ambrus for providing his valuable guidance, remarks and suggestions during the whole process.

Budapest University of Technology and Economics

Contents

Abstract

Human balancing is an important issue in many fields of everyday life, such as walking, running, cycling, carrying objects and even standing. The understanding of the balancing process is very important, especially from the point of view of elderly people. However, there are a lot of open questions about the working principle of the neural system.

We focus on the mathematical modelling of the neural process, which flows in the human neurotic system during standing still. There are several approaches in the literature. One approach is to apply a linear compensator (such as PD, PID, PIDA) in the model. Besides, model based predictive controllers are also feasible, when the human acts using a pre-learned control input pattern in the certain situations.

In our study, we compare the operation of the two control approaches, by comparing stabilometry measures, such as typical vibration frequencies, average velocity of the center of mass of the body and maximum/average tilt angle.

Összefoglalás

Az emberi egyensúlyozás az élet majdnem minden területén nagyon fontos kérdés, ilyenek például a járás, a futás, a biciklizés, tárgyak mozgatása kézzel, de még ide tartozik az egy helyben állás is. Az egyensúlyozási folyamatok megértése nagyon fontos, elsősorban az iősödő emberek szempontjából. Ennek ellenére, még mindig nagyon sok nyitott kérdés vetődik fel az idegrendszer működésével kapcsolatban.

Az idegrendszeri folyamatok matematikai modellezésére fókuszálunk, amelyek az egyhelyben állás közben zajlanak. A szakirodalomban többféle megközelítés is található. Az egyik megközelítés lineáris szabályozó (mint például a PD, PID és PIDA) alkalmazása a modellben. Emellett a modell alapú prediktív szabályozók is szóba jöhetnek, melyek alapján az ember előre megtanult mozgásmintákkal avatkozik be az egyes szituációkban.

Munkánkban a kétféle szabályozási alapelvet hasonlítjuk össze stabilometriai mérőszámok segítségével, mint például az oszcilláció tipikus frekvenciája, a súlypont mozgásának átlagos sebessége és a test maximális vagy átlagos dőlésszöge.

1 Introduction

The importance of feedback control is significantly high in the area of science and engineering. Since human-balancing is a feedback control too, the same mathematical apparatus can be used for its analysis. Balancing task is a great example to help us investigate and understand the underlying control mechanism of human brain. In spite of the huge amount of already existing scientific results, the nature and the attributes of the feedback process utilized by the central nervous system (CNS) is still a subject of discussions.

Understanding the mechanism of human balancing can result in mitigating the risk of accidental death or morbidity which derives from balance-related accidents of the elderly. While the average lifetime of Earth's population increases drastically, putting a spotlight on research in this field can be very promising in terms of saving lives. The median age of the world population is shown in Fig. 1 based on Reference [1].

Balancing might seems an easy task to do for first look, however it definitely is a complex process, which is carried out by the human brain. The brain receives many different kind of signals from the sensory organs. These signals involve pose and velocity data, acceleration, contact forces. The balance system works with the visual and the musculoskeletal systems too in order to maintain orientation. Visual signals are also conveyed to the brain about the body's position in relation to its environment are processed by the brain and compared to information from the vestibular, visual and skeletal systems. Finally, brain takes an action by sending signals to the musculoskeletal system. This results a necessary movement.

Figure 1: Median age of world population through the years

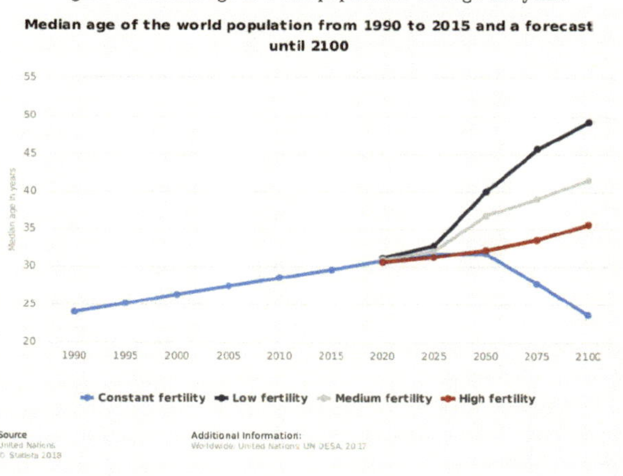

In this paper, the inverted pendulum is studied as a simple model to describe human postural balancing, similarly, as in [2]. This model is widely used in the literature, e.g. [3] together with different kind of controllers. A typical approach is to use linear compensators, such as the proportional-derivative (PD), the proportional-integrator-derivative (PID) or the proportional-integrator-derivative-acceleration (PIDA). Another approach is to use some kind of model-predictive controllers. In our study, different kind of controllers (PD and model-predictive energy-based) will be applied for the mechanical model through the process. The different controllers will be compared with measurement results of human standing still.

2 Model of human balancing

Our overall mathematical model of the postural balancing consists of the equations of motion for an inverted pendulum and the equations that formulate the control law. The model allows us to compare the operation of different type of controllers applied on the same mechanical model.

2.1 Mechanical model

The mechanical model of the human body is the inverted pendulum, which is shown in Fig. 2. This model can be taken as the model of steady postural human balancing during standing. We assume that the relative motion of the body segments are negligible. The ankle is the only joint of which the motion is considered.

The inverted pendulum is a prismatic homogeneous bar with mass m and length l. The mass m corresponds to the whole human body weight and the length l corresponds to the body height. Since the bar is homogeneous and prismatic, the gravitational force $\mathbf{G} = mg$ is acting in the geometric centre of the bar. The general coordinate of this 1 DoF system is chosen to be $\gamma(t)$. The bar is connected to the ground via a frictionless ideal pin joint. This joint corresponds to the ankle.

We model balancing as a control system, which has control inputs in general. In our model, the input torque is applied to balance the pendulum in the upright position, where γ_d is zero.

The equation of motion of the mechanical model is derived in the followings. The kinetic energy of the system is the following

$$E_{kin} = \frac{1}{2}\Theta_a\dot{\gamma}^2, \tag{1}$$

where Θ_a is mass moment of inertia of the pendulum which equals to:

$$\Theta_a = \frac{1}{3}ml^2. \tag{2}$$

The control torque is taken into account in the general force $Q = T$. The control torque does not have potential function, therefore the potential energy of the pendulum is given by the gravitational terms as follows:

$$E_{pot} = \frac{l}{2}mg\cos(\gamma). \tag{3}$$

The Lagrange equation of the second kind is as follows:

$$\frac{d}{dt}\frac{\partial E_{kin}}{\partial\dot{\gamma}} - \frac{\partial E_{kin}}{\partial\dot{\gamma}} + \frac{\partial E_{dis}}{\partial\dot{\gamma}} + \frac{\partial E_{pot}}{\partial\gamma} = Q, \tag{4}$$

where E_{dis} is the dissipative energy, which is zero in our case, since there is no mechanical damping in the system at all.

After applying Lagrange equation of second kind, our equation of motion is:

$$\frac{1}{3}ml^2\ddot{\gamma} - mg\frac{l}{2}\sin(\gamma) = T. \tag{5}$$

Based on Eq. (5), the angular acceleration is obtained:

$$\ddot{\gamma} = \frac{T + \frac{mlg\sin(\gamma)}{2}}{\Theta_a}. \tag{6}$$

Figure 2: Inverted pendulum model of balancing during standing still

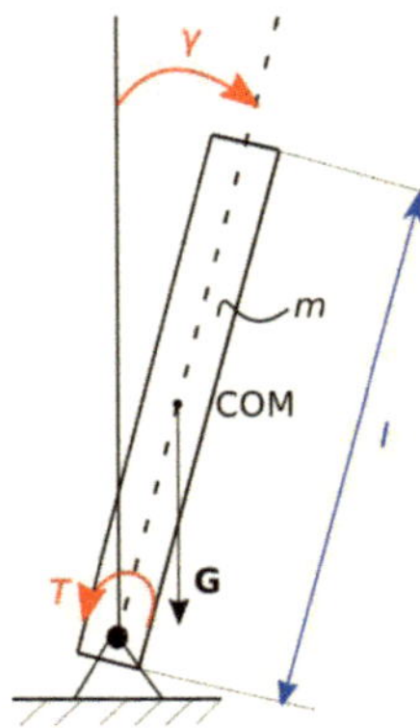

Eq. (6) is the actual mathematical model of the inverted pendulum. This model has to be augmented with the model of the controller in order to obtain the entire postural balancing model.

2.2 Controllers

Three different kind of controllers, which are possible candidates of human balancing models, are compared in this study. The first two are both proportional-derivative (PD) controllers, which are often referred as linear compensator in the literature [4]. The first controller considers the sensory dead-zone regarding the angular position measurement only, while the second one considers the dead-zone for angular velocity too. The third controller uses a concept which absolutely not uses the approach of linear compensators. Instead, the third controller considers the dynamics of the controlled plant and generates proper control impulses in order to reach the desired state. The details are given in the subsections below.

2.2.1 PD controllers

PD controllers have been widely used for the last few decades. Abbreviation PD refers to the fact that these controllers have proportional and derivative parts.

An output value that is proportional to the current error value is produced by the proportional term. The proportional response can be adjusted by multiplying the error by a constant P, which is the proportional gain constant. A high proportional gain produces a large leap in the output for a given difference in the error. If the proportional gain is too high, the system can turn out to be unstable which should be avoided. This kind of instability appears in the presence of time delay or time digitization. In contrast, a small gain results in a small output response to a large input error, and less sensitive controller. If the proportional gain is too low, the control action will not be high enough while reacting to system disturbances. Proportional gain should be most contributing one to the output, it means it should be selected properly in order to get satisfactory results.

Derivative part anticipates the system behavior and thus improves settling time and stability of the system. The time derivative of the error is computed by finding the slope of the error over time and multiplying the rate of change by D (derivative gain). All in all, PD controller is given by:

$$T = -P(\gamma - \gamma_0) - D\dot{\gamma}, \tag{7}$$

where T is the input torque which is applied to keep the pendulum from falling and γ_0 is desired angular displacement value about which angle we want to keep the pendulum tilted.

PD controller and other linear compensators, such as PID, PDA or PIDA controllers are widely used for the modelling of the neural process of human balancing [5, 6]. In ideal case, when there is no sensory dead-zone, no delay, no digital effects, these controllers make the controlled variables settle down without oscillation. However, a small amplitude oscillation is always present in case of humans. In the following Section we introduce the sensory dead-zone into the model in order to obtain human-like behaviour with oscillations that never settle down completely.

2.2.2 PD controllers with dead-zone

In general, dead-zone is a group of input values in the domain of a transfer function in a control system where the output is zero. Dead-zone regions are used to impede oscillation or repeated activation-deactivation cycles Ref. [6]. Good example for this can be air-conditioners which keep the desired temperature while having dead-zone of couple degrees.

In the case of our human balancing model, we apply a dead-zone for the angle and the angular velocity. We assume that there is a small angle or angular velocity, which is not sensible by the human sensory organs. The output signal with dead-zone is shown by blue curve in Fig. 3. This curve can be obtained as a combination of Heaviside functions.

Using Heaviside functions would drive us to deal with piecewise-smooth dynamic systems, which would highly increase the complexity of our model. In order to avoid this problem, the Heaviside function is replaced by a similar but smooth function.

Smoothing is used in our Heaviside step function in order to avoid rapid/sharp turns and noise phenomena [7]. The created approximating function tries to grasp the important patterns while gets rid of sharp edges. In smoothing, the data points of a signal are modified so individual points are diminished, and points that are lower than the adjacent points are upserged making a much smoother signal. We can set the sharpness of smoothing by changing the parameter inside. Our smooth Heaviside function is the following:

$$H(x) = 0.5(1 + \frac{x}{\sqrt{x^2 + d}}), \tag{8}$$

where, d is a parameter for tuning the sharpness and x is the independent parameter. Finally, the smoothed

Figure 3: Output torque in case of dead-zone with and without smoothing

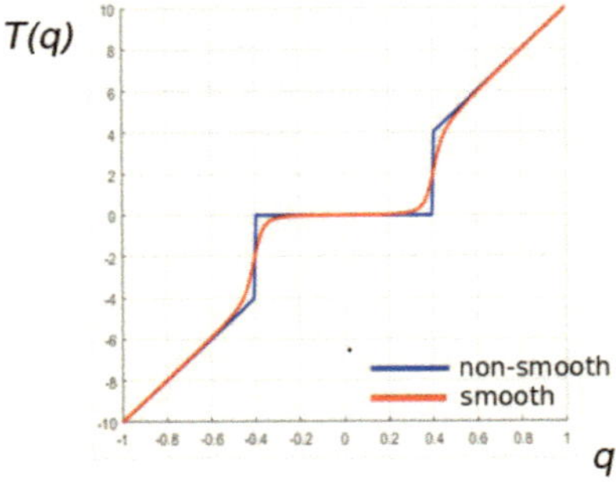

dead-zone function is composed as:

$$T_{\mathrm{d}}(x) = T(x)\left[H(x-\varepsilon) - H(x+\varepsilon) + 1\right], \tag{9}$$

which is shown by red curve in Fig. 3. Here, ε is the width of the dead-zone and $T(x)$ is the control signal without considering the dead-zone.

PD controller with a dead-zone of the angle

Here, the dead-zone for the angular position is considered only. We consider γ in dead-zone function, where γ equals to the angular displacement error relative to the upward vertical position and ε_{P} is the dead-zone width. The smoothed dead-zone function (9) is applied only for γ (angular displacement error), therefore the output torque is

$$T_{\mathrm{d}}(x) = \begin{cases} 0 & \text{if } |\gamma| \leq \varepsilon_{\mathrm{P}} \\ -P(\gamma) - D\dot{\gamma} & \text{otherwise.} \end{cases} \tag{10}$$

The numerically obtained solution by using a the PD controller with dead-zone for the angle can be seen in Fig. 4. The corresponding phase-space diagram is shown in Fig. 5. The time history of the control torque is shown in Fig. 6.

As the graphs show in Figures 4., 5. and 6., the oscillations settle down after the pendulum starts to move from a tilted position. This behaviour is not typical for humans; therefore we do not consider the application of this PD controller for modeling the neural processes. Instead, in the next section, we detail the result of a controller that considers dead-zone for the angular velocity too.

Depending on the tuning of the positive gain parameters P and D, the settling of oscillations could be faster or slower, and the overshooting at the beginning could be eliminated, but the overall behaviour would

Figure 4: Solution in time for PD controller with a dead-zone of an angle

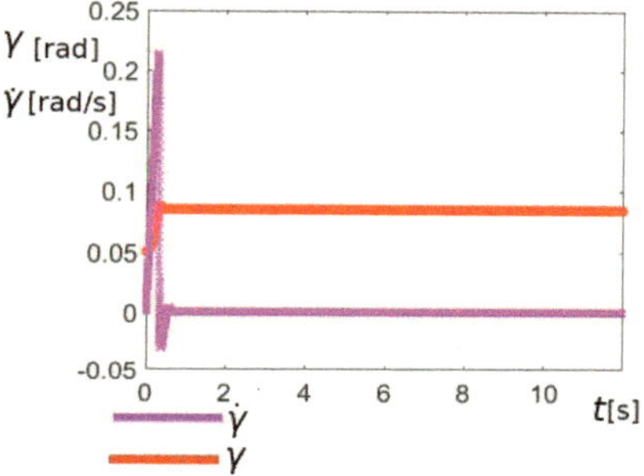

Figure 5: Phase Space diagram for PD controller with a dead-zone of an angle

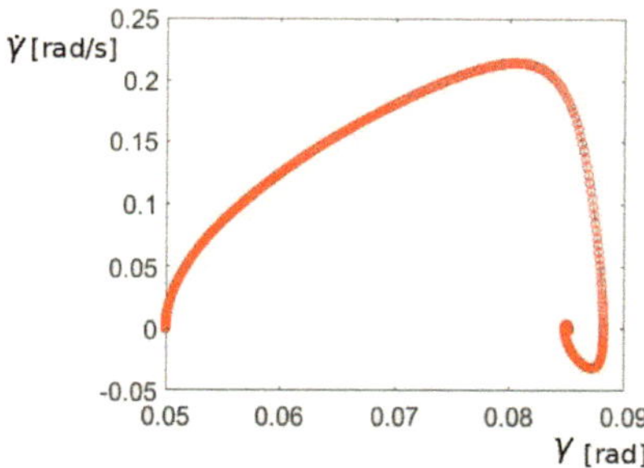

Figure 6: Torque in time for PD with dead-zone of an angle

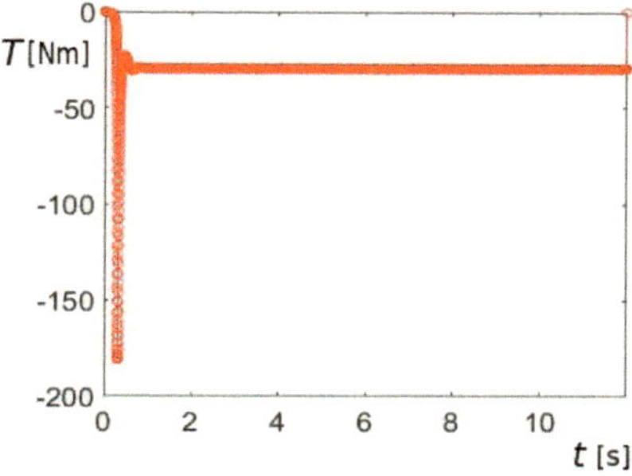

be the same: the oscillations are eliminated as simulation time goes to infinity.

PD controller with dead-zone of the angular position and the angular velocity

The smooth dead-zone function can be applied not only for proportional part of our controller but also for derivative part. Difference is that sensory dead-zone is applied for angular displacement error as well as for angular velocity of our mechanical model. So, the input torque T_{d} can be written as :

$$T_{\mathrm{d}}(x) = \begin{cases} 0, & \text{if} |\gamma| \leq \varepsilon_{\mathrm{P}} \text{ and } |\dot{\gamma}| \leq \varepsilon_{\mathrm{D}} \\ -D\dot{\gamma}, & \text{if} |\gamma| \leq \varepsilon_{\mathrm{P}} \\ -P\gamma, & \text{if} |\dot{\gamma}| \leq \varepsilon_{\mathrm{D}} \\ -P\gamma - D\dot{\gamma}, & \text{otherwise}, \end{cases} \tag{11}$$

where ε_{D} and ε_{P} are desired absolute dead-zone ranges for angular velocity and angular displacement error respectively.

The numerically obtained solution by using a the PD controller with dead-zone for the angular position and velocity can be seen in Fig. 7. The corresponding phase-space diagram is shown in Fig. 8. The time history of the control torque is shown in Fig. 9.

Figure 7: Solution in time for PD controller with a dead-zone of an angle and angular velocity

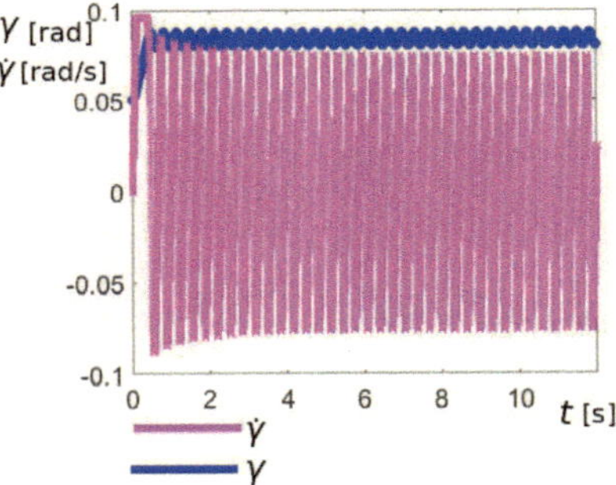

As the graphs show in Figures 7., 8. and 9., the oscillations do not settle down after the pendulum starts to move from a tilted position. Instead, after a few seconds, the oscillations have a constant amplitude. This behaviour is more typical for humans as we will see in the Section 3: some oscillation is always present in case of humans. We accept this PD controller for modeling the neural processes. However, the issue of time-delay should be considered, but it will be in future work.

Figure 8: Phase Space diagram for PD controller with a dead-zone of an angle and an angular velocity

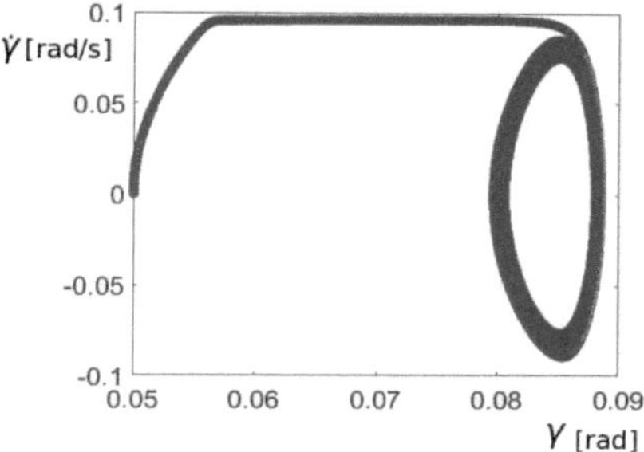

Figure 9: Torque in time for PD with dead-zone of an angle and an angular velocity

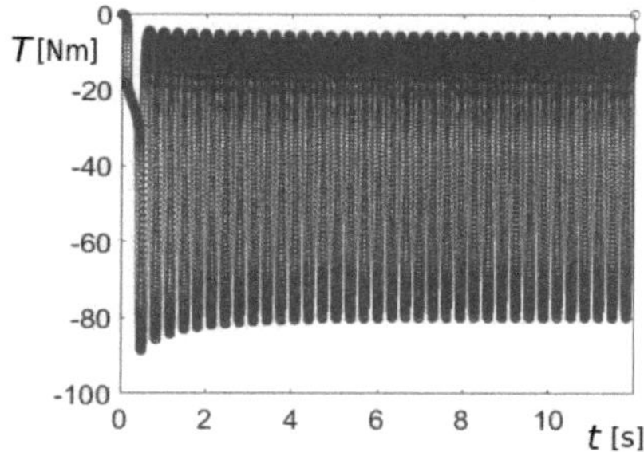

2.2.3 Model-predictive energy based controller

It is possible to consider the human brain as a model-based controller. Good example can be babies, because most of the time, they can't plan the output of their action. They learn it by trying all possible ways they can in order to perform the desired output, such as reaching an object. We can say that we build up the model of our body in our childhood. Similarly, we build the appropriate models when learning a new activity, like walking, running, skiing, etc. Postural balancing is a similar learning process and it is reasonable to try to find a model behind the neural processes.

Our model-based controller involves the issue of act-and-wait controllers. Some human movements are working with act-and-wait principle [8] because human brain is not continuously calculating all the parameters which it needs. Humans hindsight the certain events and decide the next moves based on the result which they got from the last one.

The principle of our model-predictive energy based control evaluates the potential energy U of the inverted pendulum and applies certain amplitude torque peak for certain amount of time in order to drive the pendulum near to the upward vertical position, where the potential energy U_0 is known.

The working principle of our controller tries to imitate the human behaviour: when the deviation from the vertical position becomes sensible, the brain activates a pre-learned model based action to get back to the vertical position. Once the action is activated, the brain does not send new command to the body until the end of the already initiated action.

For this controller, Phase A and Phase U will be introduced meaning actuated phase and unactuated phase respectively. When the tilt angle $|\gamma| \leq \varepsilon$, Phase U remains, but when $|\gamma|$ exceeds the dead-zone width ε, then Phase A will be switched on and torque will be exerted on the body for t_A actuation time in order to push it back to the upward vertical position.

Fig. 10 shows the prescribed time history of the actuation torque. Fig. 11 shows the Phase A and Phase U together with the pendulum.

Figure 10: Actuated and Unactuated phase for energy-based controller in time

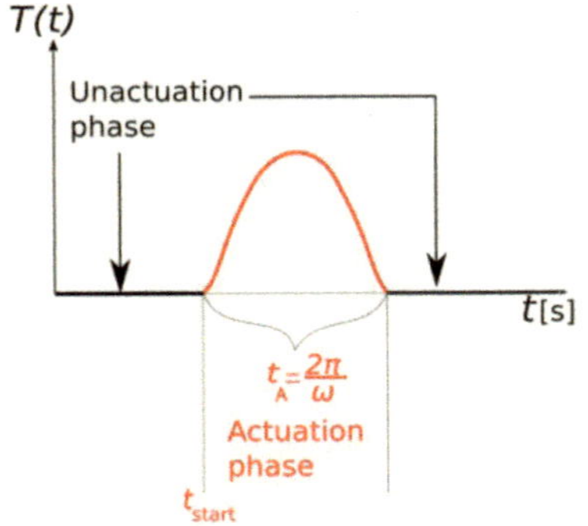

Figure 11: Actuated and Unactuated region for energy-based controller

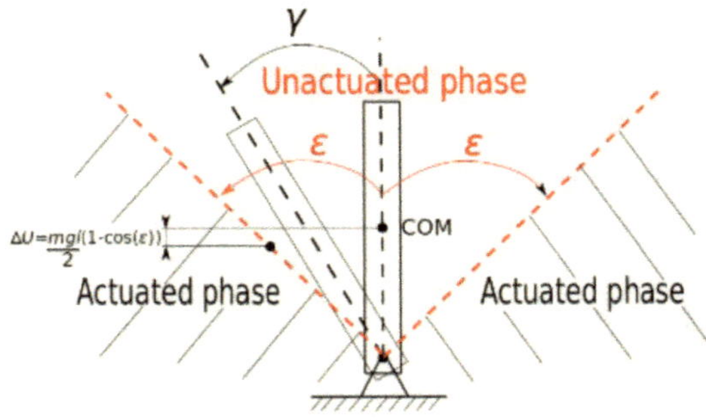

All in all, the basic idea of this control is to compensate the same amount of mechanical work which has been lost due to the potential energy loss of the falling pendulum. This control is similar to the act-and-wait control principle.

The shaping of the torque input

For the energy-based model-predictive control, a simplified model is used for our pendulum. After the linearization of the equation of motion (5) of the controlled pendulum, and after neglecting the stiffness term, the simplified equation of motion is

$$\Theta_a \ddot{\gamma} = T. \tag{12}$$

The torque action (which is activated in the Phase A) is specified by following equation:

$$T(t) = A(1 - \cos(\omega t)); \qquad t \in [t_{\text{start}}, t_{\text{start}} + t_{\text{A}}], \tag{13}$$

where A is the amplitude of the peak (its calculation will be explained later on), angular frequency ω is responsible for the width of the peak (this is an arbitrary parameter), t_{start} is the time instant of the activation of the torque and t_{A} is the time duration of the peak. The time duration t_{A} is calculated as:

$$t_{\text{A}} = \frac{2\pi}{\omega}. \tag{14}$$

The challenge here is to calculate the proper amplitude A with which the proper amount of mechanical energy is pushed into the system in case of an arbitrarily chosen ω value. In order to do this, we solve the simplified equation of motion (12) in closed form and then calculate the mechanical work W of the torque (13) over the t_{A} time duration. The work W should be equal to the potential function difference ΔU between the vertical and the actual position of the pendulum (see Fig. 11). The details are explained in the followings.

We express the angular acceleration $\ddot{\gamma}$ from Eq. (12) as

$$\ddot{\gamma} = \frac{T(t)}{\Theta_a}. \tag{15}$$

The next step is integrating the Eq. (15) in terms of time to get velocity. Here we have to assume that the torque T is constant in time. The angular velocity became:

$$\dot{\gamma} = \frac{T}{\Theta_a} t + C_1. \tag{16}$$

We need to integrate once more in terms of time to get angular displacement function:

$$\gamma = \frac{T}{\Theta_a} \frac{t^2}{2} + C_1 t + C_2. \tag{17}$$

This is the closed form solution of the simplified equation of motion (12) assuming constant control torque.

Our initial conditions are $\gamma(0) = \varepsilon$ and $\dot{\gamma}(0) = \dot{\gamma}_0$, where ε is dead-zone border value and $\dot{\gamma}_0$ is sensed velocity value at the time instance when the dead-zone border is crossed (the same time instant when we should start applying torque on the system). After using the initial conditions, the angular displacement function becomes:

$$\gamma(t) = \frac{T}{\Theta_a}\frac{t^2}{2} + \dot{\gamma}_0 t + \varepsilon. \tag{18}$$

Having the closed form solution (18), the power $P(t)$ of the control torque can be expressed. The mechanical power $P(t)$ can be written as following:

$$P(t) = T(t)\dot{\gamma}(t). \tag{19}$$

The extension of Eq. (19) is:

$$P(t) = T(t)\left(\frac{T(t)}{\Theta_a}t + \dot{\gamma}_0\right) = A(1-\cos(\omega t))\left(\frac{A(1-\cos(\omega t))}{\Theta_a}t + \dot{\gamma}_0\right). \tag{20}$$

In order to obtain the mechanical work W of the control torque, the power is integrated in time. Integrating the expression (20) in terms of time from 0 time instance to end of actuation time $t_A = \frac{2\pi}{\omega}$ results in the work achieved during this time duration

$$W = \int_0^{t_A} P(t)dt = A\left(\frac{3}{\Theta_a\omega^2}\pi^2 A + \frac{2\dot{\gamma}_0}{\omega}\pi\right). \tag{21}$$

As it is mentioned earlier, the potential energy loss U due to relative displacement of center of mass of the pendulum should be compensated by the work achieved by applying torque during actuation period. The potential energy difference will be given by following formula (see Fig. 11):

$$\Delta U = mg\frac{l}{2}(1-\cos(\varepsilon)). \tag{22}$$

So, in order to get the appropriate amplitude A value, the right hand side of equations (21) and (22) have to be equal. The equation for A can be written as below in this form:

$$A(\frac{3}{\Theta_a\omega^2}\pi^2 A + \frac{2\dot{\gamma}_0}{\omega}\pi) - mg\frac{l}{2}(1-\cos(\varepsilon)) = 0. \tag{23}$$

From Eq. (23) ,the roots are obtained as follows below:

$$A = \frac{\frac{-2\dot{\gamma}_0\pi}{\omega} \pm \sqrt{\frac{4\dot{\gamma}_0\pi^2}{\omega^2} + \frac{12\pi^2}{\Theta_a\omega^2}\frac{mgl(1-\cos(\varepsilon))}{2}}}{\frac{6\pi^2}{\Theta_a\omega^2}}. \tag{24}$$

As it is shown in Eq. (24), the amplitude A value is dependent on ε, which is an arbitrarily chosen parameter of the controller.

Simulation results with the model-predictive energy-based controller

The numerically obtained solution results by using a the model-predictive energy-based controller can be seen in Fig. 12. The corresponding phase-space diagram is shown in Fig. 13. The time history of the control torque is shown in Fig. 14 and 15. In Fig. 15, we can see that one torque peak is not enough for the stabilization of the pendulum in the upward vertical position. An extra peak always follows the original actuation signal. The inaccuracy is because of the simplifications mentioned in the previous section.

Figure 12: Solution in time for Energy based controller

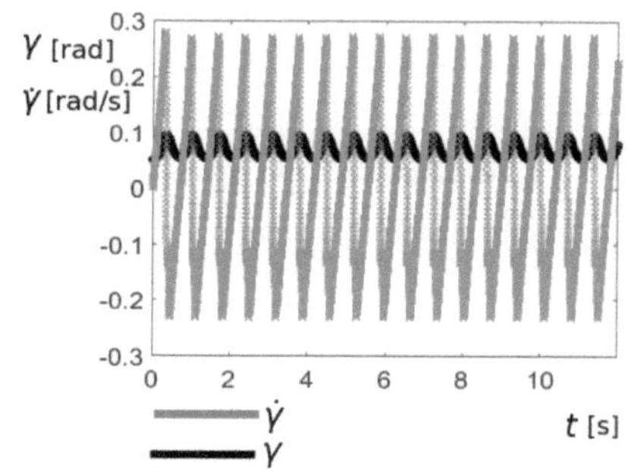

Figure 13: Phase Space diagram for Energy based controller

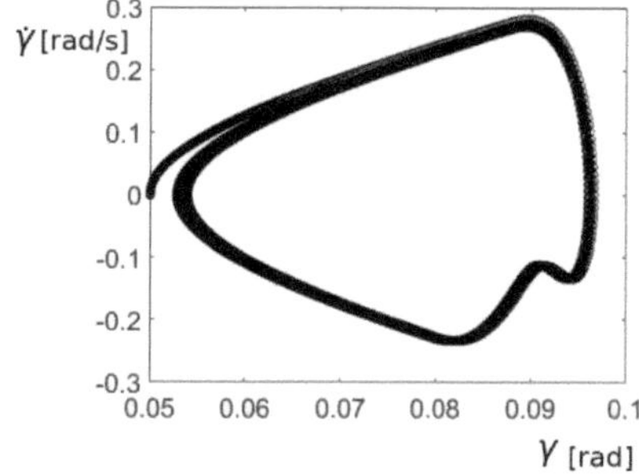

As we can see in Figures 12, 13, 14 and 15, the pendulum does not settle down in the upward vertical position. Instead, there are oscillations of which the amplitude does not increase. This seems to similar to the human behaviour.

Figure 14: Torque in time for Energy based controller

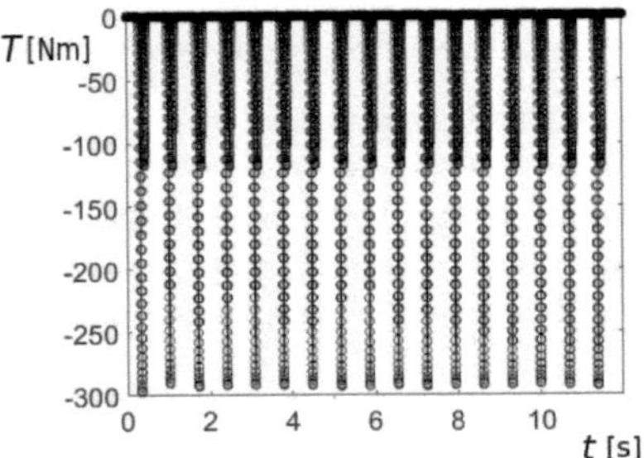

Figure 15: Torque in time for Energy based controller-closer look

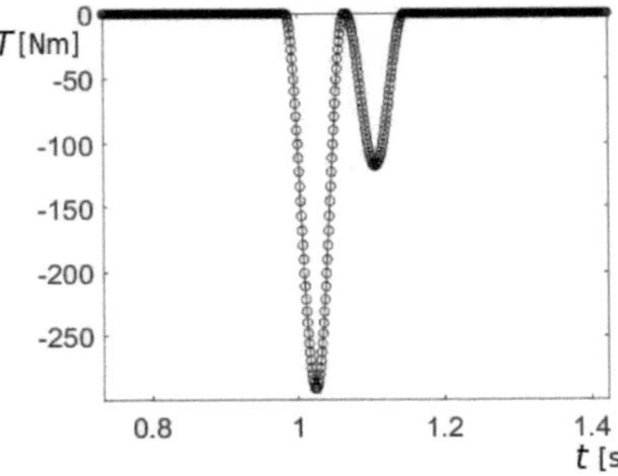

3 Measurement data and comparison with the simulations

In this section, we compare the simulation results obtained from the PD controller and the model-predictive energy-based controller with the measurement results of human standing still. The comparison was accomplished by using stabilometry values based on the references [9, 10].

3.1 Simulated and measurement data

The Figures 16, 17 and 18 show the simulation results together: the PD controller with dead-zone for angular position, the PD controller with dead-zone for angular velocity and angular position and the model.predictive energy-based controller.

As the Figures 16, 17 and 18 show, the PD controller with angular position and velocity dead-zone and the model-predictive energy-based controller show the similarity for human behaviour in the sense, that the oscillations never settle down. Therefore, only these controllers are compared to the measured human

Figure 16: Solutions in time in case of the three different controllers

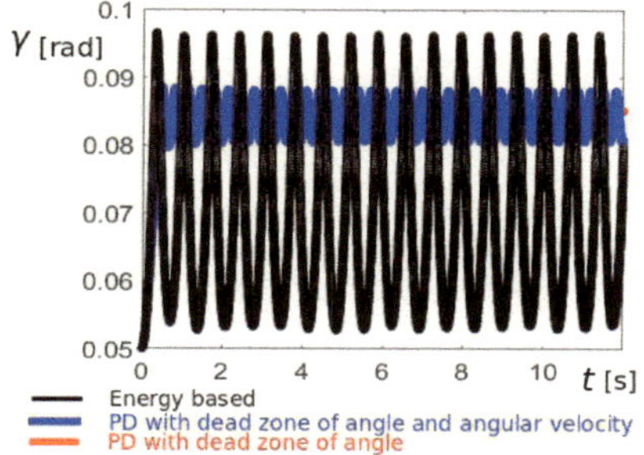

Figure 17: Phase-Space plots in case of the three different controllers

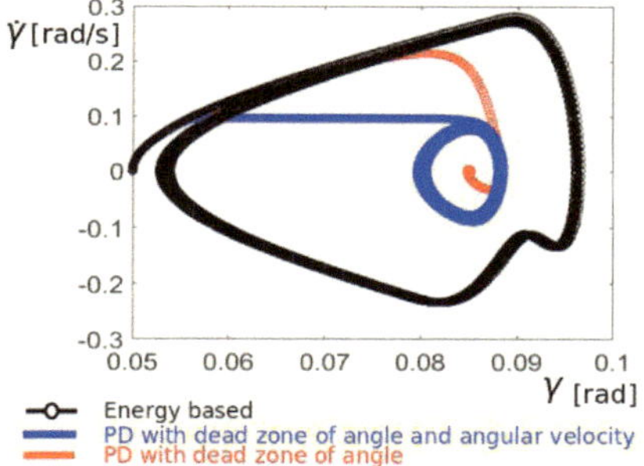

Figure 18: Time history of the control torque in case of each controllers

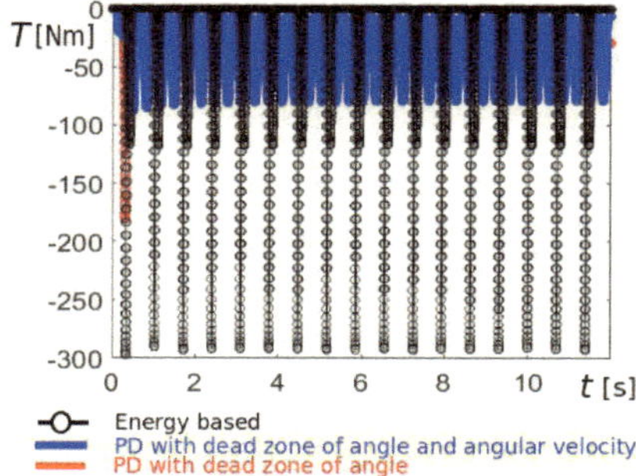

motion.

Figure 19 shows the measurement data for the human standing still. The measurement data were obtained by measuring the horizontal position of a certain point of the head. From the horizontal position, the tilt angle was possible to be calculated. The calculation of angle from displacement is given below:

$$\gamma = \arctan\left(\frac{x}{l}\right), \tag{25}$$

where x is the displacement of the head and l is the height of the person. The displacement of the head was converted to tilt angle of the body by considering the human height as 1.8 m.

For the sake of better overview, the oscillations during human standing still were measured in case of eyes open and closed. As it is visible in Fig. 19, closed eye balancing is harder task to do which leads to more deviation from zero angle.

Figure 19: Angle of sway for open and closed eye measurement

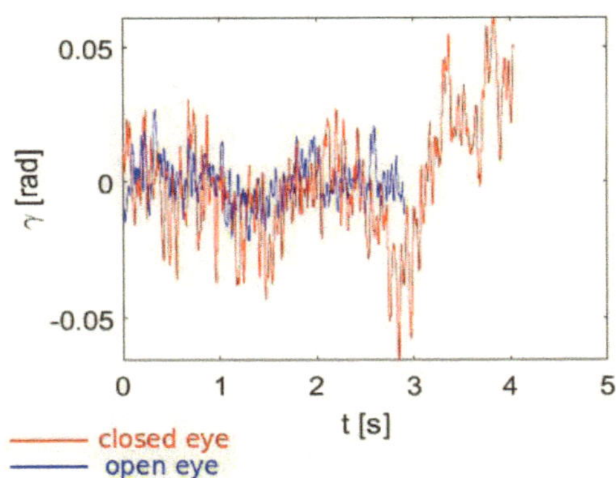

3.2 Comparison by means of the stabilometry measures

According to source [9], "stabilometry in general is the objective study of body sway during quiet standing, i.e., stance in the absence of any voluntary movements or external perturbations. Conventionally, the study focuses on the properties of body sway during upright standing, thus far primarily measured by means of force plates. Sometimes, upper body sway is studied in sitting postures. Stabilometry aims at collecting information indicative of the steady-state functioning of the postural control system, and of its success in stabilizing the body against gravity, by examining the properties of measures, directly or indirectly related with postural sway".

In our work, some of the stabilometry measures will be used in order to compare the results that we obtained from from each one of the controllers and from the measurement.

- RMS (root mean square) of the angular position (RMS of the deviation from the vertical). If this value is large, it means that the subject did not feel confident during the balancing task and the deviation from the vertical position was large.
- RMS (root mean square) of the angular velocity. Again, large values mean that there were many unwanted oscillations during the balancing task.
- frequency/FFT(fast Fourier transform) of the signal. The typical frequencies of the oscillations is an important measure of stability in case of human balancing. Higher frequencies lead us to conclude that the stability is pure.

RMS (root mean square) of the angle and angular velocity are presented from the simulation/measurement signal, which is a series of data points in time. Signal frequencies are measured by FFT (fast Fourier Transform). The fast Fourier transform (FFT) is an algorithm that sorts out a signal for a certain period of time (or space) and breaks it into its frequency components [11].

In the case of a set of n values, RMS values of the angle and the angular velocity are calculated by the formula below:

$$\gamma_{rms} = \sqrt{\frac{1}{n}(\gamma_1^2 + \gamma_2^2 + ... + \gamma_n^2)}, \tag{26}$$

$$\dot{\gamma}_{rms} = \sqrt{\frac{1}{n}(\dot{\gamma}_1^2 + \dot{\gamma}_2^2 + ... + \dot{\gamma}_n^2)}. \tag{27}$$

The RMS value of the angular position deviation for our PD controller with dead-zone of the angle and the angular velocity is:

$$\gamma_{\rm rms}^{\rm PD} = 0.0833. \tag{28}$$

The RMS value of angular velocity for our PD controller with a dead-zone of an angle and angular velocity is:

$$\dot{\gamma}_{\rm rms}^{\rm PD} = 0.0541. \tag{29}$$

The RMS values of the angular position deviation and the angular velocity are calculated for the model-predictive energy-based controller respectively:

$$\gamma_{\rm rms}^{\rm E} = 0.0705. \tag{30}$$

$$\dot{\gamma}_{\rm rms}^{\rm E} = 0.1443. \tag{31}$$

The RMS values of the angular displacement are calculated in terms of open eye and closed eye measurements respectively,

$$\gamma_{\mathrm{rms}}^{\mathrm{open}} = 0.0081, \tag{32}$$

$$\gamma_{\mathrm{rms}}^{\mathrm{closed}} = 0.0218. \tag{33}$$

The simulated signals are periodic, therefore there is no need to use FFT when the typical frequencies are calculated. The time period was determined by simply using the solution graphs. The frequency values for the PD controller and the model-predictive controller are respectively the following:

$$f^{\mathrm{PD}} = 2.9166\,\mathrm{Hz}, \tag{34}$$

$$f^{\mathrm{E}} = 1.416\,\mathrm{Hz}. \tag{35}$$

In case of the measurements, FFT was used to identify the dominant frequencies in the signals. The typical frequencies could have been identified based on the FFT (see Figures 20 and 21), but as an alternative approach, we used the 'meanfreq' command in Matlab, which provided the following frequency values:

$$f^{\mathrm{open}} = 0.22\,\mathrm{Hz}, \tag{36}$$

$$f^{\mathrm{close}} = 0.11\,\mathrm{Hz}. \tag{37}$$

As it can be seen, the highest peaks in the FFT spectrum are around the same frequency values.

Figure 20: Amplitude and frequency diagram for open eye measurement(left: whole frequency range, right: the relevant frequency range)

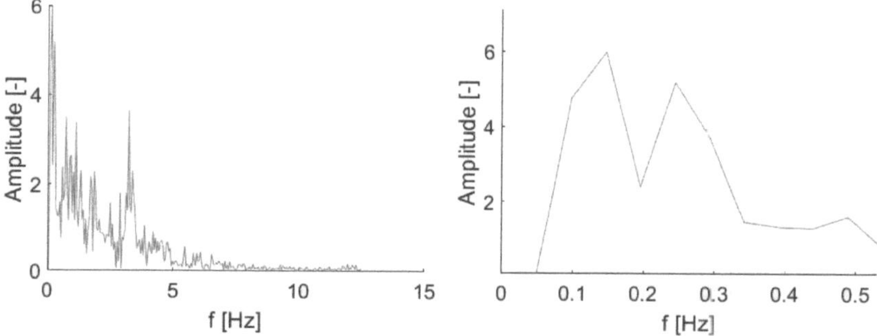

Figure 21: Amplitude and frequency diagram for closed eye measurement (left: whole frequency range, right: the relevant frequency range)

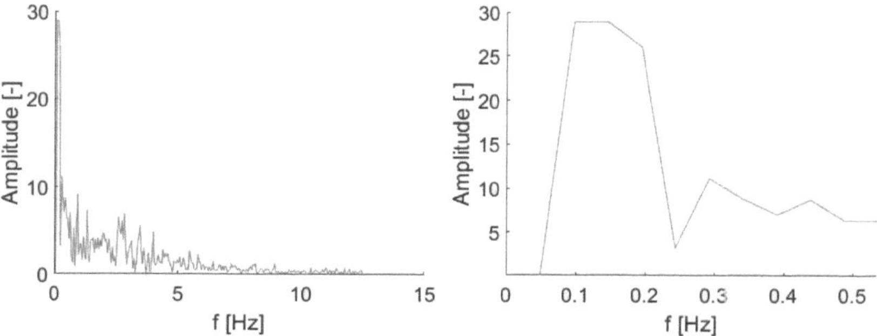

4 Results and conclusions

Plots make things clear in terms of comparing the results. According to Figures 4, 5, 6 and 17., γ and $\dot{\gamma}$ values are flattened when PD controller with angle dead-zone was applied. This is definitely not the case for humans. Hence, The non-delayed PD controller with dead-zone for the sensed angle was excluded from the possible candidates of the balancing models.

The angle γ oscillates when the non-delayed PD controller is applied with dead-zone for both the angular position and the angular velocity. The model-predictive energy-based controller also generated non-decaying oscillations. These two models were considered as possible candidates for the modeling of the neural process of postural balancing.

When we discuss the quality of the controllers, Phase-Space diagram is very important because it shows whether the stability is achieved or not. As it is shown in Fig 17, the solutions oscillates around the origin except PD controller with a dead zone of angle. As we discussed earlier, PD controller with a dead zone of an angle is not suitable for our human-like control because this controller results a desired output with almost no oscillations. Apart from that, the other controllers have stable periodic orbit, which makes a closed curve in the phase-space.

According to Fig. 6, it is worth to say that PD controller with dead-zone of angle is settling at almost constant torque value, but in reality ,it is not feasible because human is not able to apply exactly the same amount of torque for a certain period of time. That is why, this type of controller is not discussed further for comparison with other kind of controllers which resemble the human balance control.

On the other hand, PD controller with a dead-zone of an angle and an angular velocity resembles human behaviour because the control torque oscillates around a certain value. In terms of energy-based controller, it is possible to say that the resulting motion resembles the human balancing behaviour too.

As it can bee seen in Figures 7, 9 and 16, the solution and the applied torque is periodic after an initial transient in case of the PD controller. Similarly, according to Figures 12, 14 and 18, almost same amount of torque is applied with certain periodicity in case of the model-predictive controller. However, for humans completely periodic motion is not typical. Instead, humans move in an unpredictable chaotic-like manner. In order to achieve this kind of more realistic motion, we would need to apply much more complicated models, which is a possible direction of our future studies.

According to the Figures 14 and 15, for energy based controller, the torque is applied twice to compensate the work due to the falling of the pendulum. It means that one cycle of torque was not enough to compensate the work, so second cycle is applied almost immediately to keep the pendulum in utmost position. It is caused by the inaccuracies and estimations in the controller. On one hand, it would be possible to further develop the model, with which the pendulum can go to the upward vertical position after one single torque impulse. But on the other hand, we have to keep in mind, that human brain is not a computer, so the calculations made by the brain have inaccuracies too. Therefore there is no need to make the controller perfect, when the goal is the imitation of human behaviour.

Root mean square angle values of PD, energy based, open and close eye measurements are given in Eqs. (28), (30) , (32) and (33). Higher the values are, more deviating the signal is which means less sensitive control. We can conclude that the RMS value of the simulations are in the same order of magnitude as the measurements. By tuning the dead-zone width (thorough studies and literature survey are necessary for that) and the parameters of the energy based controller better fitting could be achieved in future work.

The amplitude-frequency diagrams clearly show that the amplitudes of open eye measurement is much less than those for closed eye in the whole frequency range. Figs. 20 and 21 The reason behind it is simply related to the sensor because eyes are responsible for evaluating the tilted angle between the position of the body and the horizon. When eyes are closed, evaluation gets harder because one of the important sensors is switched off. Without the eyes involved, the motion becomes more chaotic and unreliable. Typical frequency values- Eqs. (36) and (37) are valid indicators of the idea that closed eye balance measurement is more deviating in comparison with the open eye measurement. According to our results, but without drawing general conclusion, the higher typical frequency indicates that the signal is more stable. The frequencies of the simulated motions are in the range, which is typical in case of the measurements. This encourage us to use and develop further these models of balancing.

All in all, we can say that two controller models of human self-balancing were compared with the human behaviour. The model-predictive energy-based controller was developed by us, while the PD controller is widely used in the literature. These two controllers provide similar behaviour as humans; therefore we can use them in further research.

We compared our model-predictive energy-based controller with the PD controller that includes dead-zones for the angular position and the angular velocity. In case of the PD controller, the RMS values and the typical frequency values were closer to the measurements. However, we have to note that the parameters of the models can be further tuned and many factors, are not included in the model, such as more degrees of freedom or time delay. These may affect this conclusion.

References

[1] United Nations. *World population - median age.* ©Statista, 2018.

[2] Csenge A. Molnar, Ambrus Zelei, and Tamas Insperger. Estimation of human reaction time delay during balancing on balance board. In *The 13th IASTED International Conference on Biomedical Engineering (BioMed 2017)*, pages 195–199, Innsbruck, Austria, February, 20 — 21 2017.

[3] Agata Nawrocka Andrzej Kot. Modeling of human balance as an inverted pendulum. pages 254–257.

[4] Araki.M. *Control Systems,Robotics and Automation vol-2-PID control.* ©Encyclopedia of Life Support Systems(EOLSS).

[5] G. Stépán. Delay effects in the human sensory system during balancing. *Transactions of the Royal Society A*, 367(1981):1195–1212, 2009.

[6] John Milton, Tamas Insperger, Walter Cook, David Money Harris, and Gabor Stepan. Microchaos in human postural balance: Sensory dead zones and sampled time-delayed feedback. *Physical Review E*, 98(022223), 2018.

[7] Jeffrey S. Simonoff. Smoothing methods in statistics, 2nd edition. 1998.

[8] Peter Galambos Gabor Stepan Tamas Insperger, Laszlo L. Kovacs. Act-and-wait control concept for a force control process with delayed feedback. *Motion and Vibration Control*, pages 133–142, 2008.

[9] Windhorst U Binder M.D., Hirokawa N. *Encyclopedia of Neuroscience. Springer, Berlin, Heidelberg.* 2009.

[10] R. M. Kiss G. Nagymate, Zs. Orlovits. Reliability analysis of a sensitive and independent stabilometry parameter set. *Plos One*, page 14 pages, 2018. doi.org/10.1371/journal.pone.0195995.

[11] Don H.; Burrus Charles Sidney Heideman, Michael T.; Johnson. Gauss and the history of the fast fourier transform. *IEEE ASSP Magazine*, page 14–21, 1984.